FUNGI

PAMELA FOREY

CONTENTS

What are the Fungi? 2
Poisonous fungi 7
Edible fungi 11
The ecology of fungi 15
Economically important fungi 23

COVER: *The poisonous Sulphur Tuft (Hypholoma fasciculare), growing on an ivy-covered tree stump.*

Series editors: Jim Flegg and Chris Humphries

Copyright © 1985 by Pamela Forey. First published 1985.
Number 4 in the Shire Natural History series. ISBN 0 85263 746 2.
All rights reserved. No part of this publication may be reproduced or transmitted in any form or by any means, electronic or mechanical, including photocopy, recording, or any information storage and retrieval system, without permission in writing from the publishers, Shire Publications Ltd, Cromwell House, Church Street, Princes Risborough, Aylesbury, Bucks, HP17 9AJ, UK.

Set in 9 point Times roman and printed in Great Britain by C. I. Thomas & Sons (Haverfordwest) Ltd, Press Buildings, Merlins Bridge, Haverfordwest, Dyfed.

What are the Fungi?

Everyone is familiar with the toadstools which appear in the woods in the autumn, with the moulds that grow on stale bread and with the yeasts that are used to make beer. These are all members of a large group of living organisms called the Fungi. They are probably more closely related to the plants or to the algae than to the animals, but their relationship to other living things on earth is difficult to assess, since they have existed for such a long time that their origins are obscure. The earliest known fungi are fossilised in Precambrian rocks, over one thousand million years old.

The Fungi have traditionally been divided into five groups: Myxomycetes; Phycomycetes; Ascomycetes; Basidiomycetes; Deuteromycetes or Fungi Imperfecti. However, today the Phycomycetes are not considered to be a natural group since they show great diversity and are not all closely related to each other.

MYXOMYCETES

These are the slime moulds, a group of strange freeliving organisms which look like overgrown amoebae with many long thin projections or pseudopodia. They live in damp places, in and under rotting logs, in the soil and in woodland leaf litter, growing gradually larger and reproducing by simple division, fragmentation, or by the production of spores. The slime moulds are usually separated from the other fungi (some zoologists have even classed some of them as animals) into a group known as the Myxomycota.

All the other fungi are then grouped together into the Eumycota or true Fungi. They are usually formed of long threads called hyphae, growing in a tangled mass (or mycelium) in the simple moulds or mildews, or twined closely together to form more solid structures as in the toadstools or orange peel fungi. Fungi do not have chlorophyll like green plants do and so cannot synthesise their own food. Instead, they feed as saprophytes on dead and decaying materials or as parasites growing in live hosts, invading live tissues and often causing great damage. The fungi reproduce by spores, produced asexually by simple cell division or sexually by the union of two cells. The two types of spores are produced quite separately and at different stages in the life cycles of the fungi, which may be very complicated as a result (see fig. 14). The spores and the sporangia in which they are produced are very distinctive and provide a basis for the classification of the fungi.

PHYCOMYCETES

These are the moulds, the simplest fungi, consisting of single cells or, more often, of tangled masses of hyphae — a mycelium — growing, for example, on over-ripe fruit or bread. It is the smallest group of fungi, with about eleven hundred known species, a very mixed group consisting of the water moulds, the downy mildews and the bread moulds, amongst others. Some of them reproduce asexually by zoospores, tiny mobile spores which swim through the thin films of water in the soil or in ponds, and sexually by oospores, thick-walled resistant spores formed by the union of the contents of two cells. One such mould is *Pythium debaryanum*, which lives in the soil and feeds saprophytically on dead leaves and other organic remains or parasitically, when it may cause problems in the garden, especially in greenhouses and cold frames, attacking seedlings and causing Damping-off Disease. It grows by means of long white hyphae, rapidly invading seedlings, often where the base of the stem or the roots have been injured, during pricking out for example, and spreading by means of zoospores, which swim through the soil water to other seedlings. The fungus also produces oospores, resistant spores with a thick black spiky coat, formed sexually from the fusion of an egg cell and a male cell; these can be blown by the wind to other areas and new sources of food or can survive the cold months of winter.

Many members of the Phycomycetes follow this pattern of growth and local spread by hyphae and zoospores, then formation of sexual resistant spores in the autumn or for dispersal. Examples in-

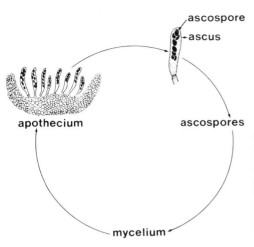

A

Fig. 1. *Life cycle of an ascomycete fungus, like Otidea onotica (plate 1), showing apothecium with asi and ascospores (see also fig. 3). The mycelium of such a fungus is rarely visible, being buried in soil, in leaf litter or in the wood of a rotting log.*

clude *Saprolegnia*, a genus of water moulds which grow on dead fish and attack their eggs; *Plasmopara,* a genus of downy mildews including *P. viticola*, the downy mildew of grapes, which has been known to cause considerable damage in vineyards; and *Synchytrium endobioticum*, which causes Wart Disease on potatoes.

Other phycomycetes reproduce asexually in rather different ways. The Black Bread Mould, *Rhizopus stolonifera*, produces round sporangia (spore-producing structures) on upright hyphae, in which hundreds of spores are formed, to be released and carried away in air currents when the sporangia burst. The oospores are similar to those in *Pythium*. Phycomycetes with similar kinds of asexual reproduction include *Pilobolus*, a genus of moulds found on dung (fig. 12), and *Entomophthora muscae*, which is parasitic on flies.

ASCOMYCETES

The largest group of fungi are the Ascomycetes, with some thirty thousand species, eighteen thousand of which are associated with algae to form lichens. The other twelve thousand 'freeliving' members of the Ascomycetes are a varied collection of fungi, ranging from the one-celled yeasts to the mould-like *Penicillium* and complex organisms like the truffles and morels.

Many members of the group reproduce asexually by conidiospores: these are formed on upright hyphae as strings or clusters of single cells. But especially characteristic of the group is the ascus, a specialised sporangium inside which are formed eight (or multiples of eight) ascospores. This represents the sexual reproductive phase of an ascomycete fungus, for each ascospore is formed from the fusion of two special cells, equivalent to sex cells. The asci are often produced inside special fruiting bodies, ball-like cleistothecia with one or a few asci inside, flask-shaped perithecia with a single opening at the top, or cup-shaped apothecia lined on the inside with many asci (see fig. 1).

The simplest ascomycetes are the yeasts, which consist of many single cells that can multiply very rapidly by budding off a new cell from each old one when food and water are abundant; large clusters of cells may result. If the food or water supply becomes scarce then asci are formed — larger cells inside each of which are produced four resistant ascospores. Yeasts are common in the wild, growing in the nectar of flowers and on ripe fruits, where sugar is present in large quantities, and forming ascospores in the autumn which can remain in the soil until the following spring. Yeasts are economically important and used in fermentation processes in the making of beer and wine and in breadmaking.

Many of the ascomycete fungi have a superficial resemblance to the phycomycete moulds and mildews. Species of *Penicillium* and *Aspergillus*, growing on decaying vegetables and stored foodstuffs, at first form a tangled web of whitish hyphae over the surface of the food. Soon, however, they begin to form masses of blue, green or yellow powdery conidiospores, which transform the moulds, giving them the name of powdery mildews and enabling them to spread very rapidly. Although many of the powdery mildews are saprophytes, others are parasites, including several of economic importance like *Sphaerotheca macularis*, the Hop Mildew, and *Erysiphe polygoni*, the Pea Mildew, both of which can cause rapid devastation of a crop if not caught early. The asci of these fungi are rarely produced and the fungi rely on conidiospores for dispersal. When asci are formed, they are enclosed in ball-shaped cleistothecia, which blow away in the wind and split open to release the spores.

In meadows and woodlands grow many larger ascomycetes, resembling small toadstools of the Basidiomycetes in their size, rather than the moulds or mildews. Several groups of these form cup-like apothecia, like those of *Otidea* (plate 1) and *Peziza*, which resemble pieces of brown, orange or yellow orange peel; in others, like *Helvella* (plate 2) and *Morchella* (plate 3), the cup-shaped fruiting bodies are folded and turned inside out and borne on thick stalks.

BASIDIOMYCETES

The Basidiomycetes are the most familiar group of all, with about 13,500 species and including toadstools, bracket fungi, club fungi and puffballs, amongst others. These are all fruiting bodies of the fungi, on which are produced large numbers of specialised sporangia called basidia (see fig. 2). On the outside of each basidium four basidiospores are formed. This represents the sexual reproductive phase of a basidiomycete fungus, for each basidiospore is formed from the fusion of two special cells, equivalent to sex cells, followed by immediate reduction division. Very few of these fungi reproduce asexually but basidiospores are produced in huge numbers.

Toadstools are the fruiting bodies of the largest group of basidiomycetes; they are often the only visible portion of the fungus: the mycelium is hidden amongst dead leaves in the soil, in the wood of rotting logs or in the bark of trees. Most toadstools consist of a stalk and cap, on which are many gills radiating outwards from the centre to the edges of the cap

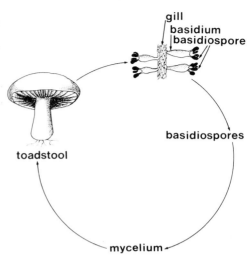

Fig. 2. *Life cycle of a basidiomycete fungus, such as Laccaria amethystea (plate 5), showing toadstool with gills. An enlargement of one gill shows how the basidia and basidiospores are formed (see also fig. 4).*

(plates 4 and 5). Each gill is covered with thousands of basidia, each budding off four basidiospores, millions of which are formed (fig. 2). If the cap of a toadstool is placed on a piece of paper overnight, a spore print is formed by these myriads of spores and shows the exact pattern of the gills. The toadstools of the genus *Boletus* and its relatives are rather different for, like bracket fungi, the undersides of their caps are covered with pores instead of gills.

Bracket fungi can be seen in woodland areas all over Britain and Europe, particularly on dead and dying trees. They are the fruiting bodies of fungi whose mycelium is hidden beneath the bark of the tree; many are annual, like the *Polyporus* and *Piptoporus* species (plate 6), in the sense that new fruiting bodies appear each year but a few, like *Fomes annosus*, have perennial fruiting bodies which form a new ring on the outer circumference each year. The underside of the bracket fruiting body is covered with small pores, each of which is the opening of a vertical tube lined with thousands of basidia. The

Plate 1. *The cup-shaped fruiting bodies (apothecia) of the ascomycete fungus Otidea onotica are found growing in clusters in leaf litter of deciduous woodland during autumn. They are rather elongated lop-sided cups with ragged margins, up to 6 cm long and 10 cm deep, yellow in colour with pinkish streaks. This is often known as the Lemon Peel Fungus or Hare's Ear.*

Plate 2. *Saddle Fungus or False Morel (Helvella crispa). Found in deciduous woodland in late summer and autumn, it grows in clearings and along grassy paths. It is usually white with a thick, deeply grooved stalk bearing a thin-lobed folded cap. Spores are produced on the upper surface of the cap and are expelled in tiny 'puffs' which may be visible on a still day.*

Plate 3. *Morels are the fruiting bodies of the ascomycete Morchella esculenta. They grow up to 20 cm high, in woodlands and gardens in the spring, especially on burned ground or a bonfire site. They have an aromatic, spicy taste and can be eaten fresh but not raw. They are often dried and then stored as a spice.*

basidiospores are ejected from the basidia into the centre of the tube, where they fall down and out of the pore to be carried away by the wind. The whole process results from a miracle of precision engineering for the tubes are very narrow; they may be several centimetres long and they must be exactly vertical or the spores would catch on the sides as they fell. Since the bracket is growing at an angle to the tree trunk that in turn is growing at another angle, the vertical orientation of the tubes is amazing, controlled by precise responses of the fungus to the force of gravity. The number of spores is also startling — a large fruiting body of *Ganoderma*, for example, may release twenty thousand spores per minute and goes on doing so for up to five months of the year.

Puffballs and stinkhorns belong to another group of Basidiomycetes which form egg-like fruiting bodies at ground level, from a mycelium in the soil. In puffballs the basidia line the inside of the 'egg' and eventually break down to release hundreds of thousands of powdery spores. The outer coat of the puffball becomes dry and papery and an opening appears at the top; raindrops falling on the puffball, the touch of the wind passing across the opening and insects crawling over the top all create pressure on the thin coat and force the spores to 'puff' out like talcum powder (plate 7). In stinkhorns the 'egg' splits open and an erect stalk expands upwards, reaching its full height in under two hours and topped by a greenish-black crinkled dome-shaped cap, which soon partially disintegrates to form a slimy mass of basidiospores, stinking of rotting flesh (plate 8). The smell is so strong that it may disturb the people living nearby, and it attracts swarms of flies, which feed on the sticky mass and carry away the spores on their feet.

Not all basidiomycetes have such complex fruiting bodies. Some, like the Jew's Ear Fungus *(Auricularia auricula*, plate 9), and the Brain Fungus *(Tremella mesenterica*, plate 10), have simple gelatinous fruiting bodies in which the basidia are embedded. The basidiospores are ejected forcibly when ripe, away from the fruiting bodies and into the currents of

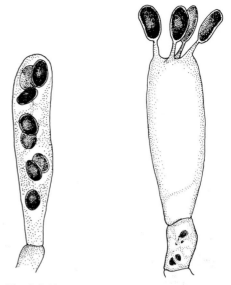

Fig. 3 (left). *Ascus of an ascomycete, with eight ascospores inside.*

Fig. 4 (right). *Basidium of a basidiomycete, with four basidiospores formed externally.*

air, to be carried away by the wind.

Rusts and smuts are parasitic basidiomycetes which can cause great devastation to many species of plants, including garden and crop species. Rusts often have complex life cycles, with several kinds of asexual spores formed in succession as well as basidiospores, and more than one kind of host (fig. 14). Economically, the most important of these fungi is probably the Black Rust of Wheat, *Puccinia graminis*, with which agriculturalists have been fighting continuously for many years. Amongst the smuts are several species which attack wheat, oats and other cereals and grasses. The life cycles of smuts are generally much simpler that those of rusts, with fewer kinds of asexual spores formed and only one host (fig. 13).

DEUTEROMYCETES OR FUNGI IMPERFECTI

About eleven thousand species of fungi are known from their asexual stages only, reproducing by conidiospores and lacking oospores, ascospores or basidiospores. They are impossible to classify accurately and are therefore included in the imperfect fungi, but it has never been suggested that this is a natural group, only a convenience for fungi whose true relationships are impossible to evaluate. Amongst the imperfect fungi are species like *Botrytus cinerea*, the Grey Mould, which attacks soft fruit, and *Dactylella bembicodes*, which lives in the soil, where it traps nematodes. The conidia of most Fungi Imperfecti are similar to those found in many ascomycetes and it is probable that many of these fungi actually belong to the latter group. Occasionally one of them is found with a perithecium or apothecium and then accurate classification is possible.

Poisonous fungi

Fungi have a reputation for poisons that is out of all proportion to reality. There are more poisonous flowering plants than there are poisonous fungi, in proportion to their respective numbers. The deadly reputation of the fungi lies almost entirely in the fact that some of them can indeed cause death. The mortality rate of the Death Cap *(Amanita phalloides,* plate 11) was as high as 90 per cent before scientists worked out how the toxins (called amatoxins) affected the body and even now the mortality rate is often over 50 per cent.

Initial symptoms of Death Cap poisoning take the form of abdominal cramps, vomiting and diarrhoea, but they do not appear for up to twenty-four hours after the toadstools are eaten, by which time the toxins have usually caused irreversible liver and kidney damage, leading to death within a few days. There are several toxins present, all but one of which are unaffected by cooking. One toxin causes the gastric symptoms while the others lower the glucose and salt levels of the blood and cause extensive damage to the liver and kidneys. If the blood imbalance is corrected by intravenous glucose and saline drips, then the liver and kidney damage is considerably reduced, but early diagnosis is essential and treatment is best begun before the gastric upset reveals the poisoning.

Death Cap is only one member of the genus *Amanita,* many of which are poisonous to a greater or lesser extent, and several of which are amongst the most common of the British toadstools. They can all be distinguished by the volva at the base of the stalk, remnant of a membrane that covered the fruiting body at the button stage, and remaining as a rim in the mature toadstool of the Blusher *(Amanita rubescens),* as a distinct cup in the Panther Cap *(A. pantherina),* or as a sac-like structure in the Death Cap. In the Fly Agaric *(A. muscaria,* plate 4) the volva remains as shaggy tufts on the stalk. In northern parts of Europe this toadstool is used as an intoxicant: it causes dizziness, stupor and hallucinations followed by nervous excitation. One of the problems with using this toadstool is that it varies considerably in its potency, depending on the content of toxin present. This is most concentrated in the skin and also varies with the growth stage of the toadstool and the time of year. Dried toadstools contain another, more potent, hallucinogen, which has been known to kill if consumed in quantity.

Several species of toadstools, other than the *Amanitas,* can prove fatal. Of these, *Cortinarius speciosus* contains toxins which cause kidney damage; *Inocybe patouillardii* causes blindness, dizziness and a lowering of the body temperature and can cause death; *Entoloma lividum* causes violent sickness or death and *Gyromitra esculenta,* a morel-like ascomycete, contains a toxin which dissolves the red corpuscles of the blood. In this last species and in many other potentially poisonous species of fungi, the toxins are destroyed by heat. *Amanita rubescens,* the Blusher, for instance, can be eaten safely after cooking but not raw, as can several species of *Russula* and *Lactarius*. Many other fungus species are not actually poisonous but are highly indigestible — the human digestive system is simply not designed to cope with them. These include the Yellow-staining

Plate 4. *The Fly Agaric (Amanita muscaria)* gets its name from its use in medieval times, when it was crushed into dishes of milk to stupefy flies which came to feed on the milk. It is one of the easiest toadstools to recognise, with its red white-spotted caps, growing up to 10 cm across, which appear in late summer and autumn beneath birch and pine trees.

Plate 5. *The Amethyst Deceiver (Laccaria amethystea)* is a relatively small but distinctive violet-coloured toadstool that grows in damp shady woodland, often with beech trees, in the late summer and autumn. It has a flattened cap, up to 6 cm across, with large, broadly separated gills dusted with spiny white spores.

Plate 6. *Birch Bracket (Piptoporus betulinus)* growing on a Silver Birch tree, together with Cramp Balls *(Daldinia concentrica)*. The bracket fungus forms an annual fruiting body which persists for up to a year; it grows up to 20 cm across and 6 cm thick, smooth and white at first but darkening with age. Cramp Balls are the shiny black fruiting bodies of an ascomycete fungus which lives on ash, birch and beech trees. They appear in the autumn, grow to about 10 cm in diameter by the spring and then release thousands of ascospores from perithecia which cover the surface.

Plate 7 (top). *Clusters of yellowish white puff-balls of Lycoperdon perlatum. A relatively common sight on downs, heaths and in woods in late summer and autumn, they grow up to 9 cm across and have a cheesy consistency when young. As they ripen they become thin and papery and an opening appears at the top through which the spores escape.*

Plate 8 (above). *The Common Stinkhorn or Wood Witch, the fruiting body of Phallus impudicus, appears in late summer and autumn, growing in the leaf litter of woodlands and gardens. It may grow up to 20 cm in height.*

Plate 9 (right). *The rubbery Jew's Ear Fungus (Auricularia auricula-judae) can be recognised as a basidiomycete fruiting body under a microscope, for it has many basidia embedded in its surface. It grows up to 8 cm across, on living and dead trees, especially on elder.*

Mushroom *(Agaricus xanthodermus)*, *Boletus erythropus* and *Lactarius piperatus*. Other toadstools, like *Paxillus involutus*, the Brown Roll-rim (plate 12), may cause allergic reactions in some people but not in others, so that they acquire a mixed reputation, for edibility or for being poisonous, depending on the experience of the person involved. One species, *Coprinus atramentarius*, the Common Ink Cap, is perfectly edible provided it is eaten without alcohol. The toxin present dissolves in the alcohol and can then be absorbed by the human body, when it causes nausea and heart palpitations in a similar way to the drug Antabuse, which is used to help alcoholics refuse alcohol.

One of the most dangerously poisonous fungi, with a long history of causing illness and death, is the ascomycete Ergot of Rye *(Claviceps purpurea*, fig. 5). Its ascospores infect wild rye plants and cultivated varieties of rye in the spring, germinating within the flower and converting the flower ovary into a hard purple structure, the ergot, by harvest time. In wild rye the ergots fall to the ground and overwinter, producing many flask-shaped perithecia in the spring and forming a reservoir of spores from which the fields of rye are infected. This disease was once common in central Europe, where cultivated rye has been grown for centuries to produce rye bread. If the purple ergots are harvested and milled together with the grain, then the resulting flour is contaminated or ergotised. Consumption of bread made with this flour results in an attack of acute ergotism with abdominal pain and burning sensations in the arms and legs, followed by hallucinations, convulsions and even death. Long-term consumption of slightly ergotised bread leads to a dry painful gangrene of the extremities and people may lose hands and feet, noses and ears before death finally ensues. This condition is known as St Anthony's Fire and is caused by several different toxic alkaloids present in the ergots. In minute quantities the alkaloids are useful in medicine, particularly in childbirth since they stimulate labour, a property known and used by midwives for centuries. In several central European countries today, artifi-

Fig. 5. *Ergot of Rye (Claviceps purpurea) on wild rye grass.*

cial infection of cultivated rye is carried out and the ergots are harvested for use in medicine.

Much more difficult to recognise as dangerous are some of the moulds of the genus *Aspergillus*. *A. flavus*, for instance, is a typical storage mould, one that infects nuts and grains in storage, often after initial contamination at harvest. Under certain conditions, if the temperature is over 12 Celsius (54 Fahrenheit) and the air is humid, and on some foods more than others, the mould may spread rapidly, producing many yellow conidiospores. It also produces lethal amounts of a toxin called aflatoxin, especially if it is growing on peanuts. These are used as a feedstuff for poultry in many countries, and birds fed on infected peanuts have been known to suffer massive mortality from liver damage and internal haemorrhage. In some parts of the world peanuts are an important source of protein and calories in the human diet. The people who eat these peanuts may consume significant amounts of aflatoxin from contaminated peanuts. One of the most serious symptoms of aflatoxin poisoning is primary cancer of the liver; the incidence of this lethal disease in these countries is considerably higher than in countries where peanuts are eaten in small quantities, or than in countries, like the United States, where stored peanuts are checked regularly.

Edible fungi

Fortunately not all fungi are poisonous. Some of them, like truffles and morels, are considered to be great delicacies and command high prices on the world market. For most people, however, the common mushroom is much more familiar. Two species of mushrooms of the genus *Agaricus* are grown commercially, *A. campestris*, the Cultivated or Field Mushroom (fig. 6), and the larger Horse Mushroom, *A. arvensis*.

It is not feasible for the average mushroom lover to grow his own mushrooms — the whole procedure is too fraught with difficulties and failure can occur at many points. But wild mushrooms can still be found in fields and meadows. It is essential, however, that the many toadstools be identified accurately for if one of the poisonous *Amanita* species is mistaken for a mushroom and eaten then the consumer could die within a week.

Toadstools (of which mushrooms are just a few kinds) are identified by their colour, texture, scent and taste; by the colour of their flesh when cut; by the way the gills are formed and whether they are brittle or soft; whether the cap leaves a ring on the stalk as it opens (e.g. *Agaricus* species); whether there is a volva at the base of the stalk (e.g. *Amanita* species); and finally by the colour of the spores. They also have characteristic ecological features, like the way in which they grow — singly (e.g. *Russula* species), in clumps (e.g. *Hypholoma* species) or in rings (e.g. *Marasmius oreades);* and the habitats in which they are found — meadows or woodland, in rotting wood or parasitic on trees; and by the time of year at which they appear.

Field Mushrooms grow in clumps in meadows and pastures, together with the Horse Mushrooms, which may form rings like the Fairy Ring Toadstool *(Marasmius oreades)*. The caps of the Field Mushroom are rarely more than 10 centimetres (4 inches) across, in contrast to those of the Horse Mushroom, which may reach up to 20 centimetres (8 inches) across. They can also be distinguished by their scents for while the Field Mushroom has a typical mushroomy smell, the Horse Mushroom often smells of aniseed.

Mushrooms are not the only edible species growing in the woods and pastures in the autumn. Most of the *Agaricus* species are edible, although those staining bright yellow when cut should be avoided. Other edible species include the Penny Bun *(Boletus edulis,* plate 13), the Chanterelle *(Cantharellus cibarius,* fig. 8), the Parasol Mushroom *(Lepiota procera,* plate 14) and the Wood Blewit *(Lepista nuda,* formerly *Tricholoma nudum,* fig. 7). There are many *Boletus* species growing in Britain in deciduous and coniferous woodlands, all readily

Fig. 6. *Field Mushrooms (Agaricus campestris) have creamy white caps with deep pink gills which become dark brown with age and produce dark brown spores. The caps are up to 10 cm across and firm and smooth in texture with a mushroomy scent.*

Plate 10. *As its name suggests, the fruiting body of the Yellow Brain Fungus (Tremella mesenterica) resembles a yellow brain, with soft gelatinous folds. It grows up to 10 cm across, on dead wood, stumps, logs and twigs throughout the year, although it is more common in the autumn and winter.*

Plate 11. *The Death Cap (Amanita phalloides) grows in mixed deciduous woodland, particularly in association with oak trees, appearing in late summer and autumn. The most deadly of the fungi, death has been known to result from consumption of just one spoonful of toadstool. It is pale greenish yellow or white with a pale brown cap, up to 12 cm across, and has a membranous sac-like volva at the base of the stalk, with remnants of this membrane on the cap and a similar ring beneath the cap. It has a sickly sweet scent.*

Plate 12. *The toadstools of Paxillus involutus, the Brown Roll-rim, grow as single specimens, with caps up to 12 cm across, often in large scattered patches in birch and pine woods, in late summer and autumn. Though once considered to be edible, it now appears that some people are allergic to it. For this reason it is probably best avoided. If it is eaten it needs to be cooked for some time, as in a stew for instance.*

Plate 13. *In continental Europe the Cep or Penny Bun (Boletus edulis) is the most popular of the edible toadstools, accounting for almost half of all the fungi eaten. They can be cooked in butter or as part of a ratatouille, stuffed with sausage meat and baked, cooked in cream or used to flavour stews and casseroles.*

Plate 14. *Unlike most toadstools, the Parasol Mushroom (Lepiota procera) will not grow in shady woodland areas but prefers open grassland and sunny woodland clearings. It is edible and relatively easy to identify, with a large buff-coloured shaggy cap, up to 25 cm across, and a scaly stalk. It can be distinguished from Lepiota rhacodes, a similar but smaller toadstool which can cause gastric upsets, for this species has a smooth stalk with no scales.*

Plate 15. *Toadstools of the Rufous Milkcap (Lactarius rufus) are common in late summer and autumn, growing under pine trees, especially where the soil is rather wet and peaty. Like many other milkcaps, they have rather funnel-shaped caps when fully mature, up to 10 cm across in this species, and exude drops of fluid or 'milk' when damaged. The fluid varies considerably in colour and flavour from species to species, being white and particularly hot and peppery in the Rufous Milkcap.*

Fig. 7 (left). *During autumn and early winter the toadstools of the Wood Blewit (Lepista nuda, formerly Tricholoma nudum) grow in woodlands and hedgerows, where they may form rings, and in gardens, where they are often found on compost heaps. The caps grow up to 12 cm across, pale lilac when young, changing to pale brown as they age but retaining the lilac colour of the gills, with pale pink spores and a fragrant scent.*

Fig. 8 (centre). *The Chanterelle (Cantharellus cibarius) is a bright yellow toadstool, with a convex or funnel-shaped cap up to 10 cm across and a faint scent of apricots. It grows in clumps in mixed deciduous and coniferous woodland in summer and autumn.*

Fig. 9 (right). *The Horn of Plenty (Craterellus cornucopioides) is dark brown and leathery in texture with a funnel-shaped or horn-shaped cap, up to 8 cm across. It often forms large clumps in late summer and autumn in the leaf litter of deciduous woods, especially under beech trees.*

identified by the fact that they have pores instead of gills beneath the cap. Not all are edible but the Penny Bun can sometimes be identified by a bun mark on its brown cap, by the white line around the outside of the cap, which reaches up to 25 centimetres (10 inches) across, and by the greasy feel of the cap especially in wet weather; the stalk is bulbous and covered with a network of fine white lines.

The members of the genus *Cantharellus*, to which the edible Chanterelle belongs, have thick folds instead of gills running some way down the stalk from the underside of the convex or funnel-shaped cap and a wavy outline to the edge of the cap. Although they are edible not all are recommended for eating; however, the yellow Chanterelle *(C. cibarius*, fig. 8) and its relative the Horn of Plenty *(Craterellus cornucopioides*, fig. 9) are considered excellent when young and fresh. The Horn of Plenty has no gills, but the upper part of its stalk may be rather wrinkled. It is important to distinguish between the Chanterelle and the rather similar toadstools of the genus *Clitocybe*, at least two of which are deadly poisonous although others are edible. These fungi are much paler in colour, mostly cream or off-white, and they have true gills.

The similarity between edible and poisonous fungi is one reason why the would-be collector has to be able to identify not just the genus but individual species accurately. It is not possible to say that this genus is edible or that one poisonous: a genus often contains both deadly and edible species, which may resemble each other quite closely. There are many tales about how to distinguish poisonous from edible species — mostly old wives' tales with little basis in fact. For instance, the fact that a toadstool peels is not an accurate guide to edibility: the most poisonous one of all, the Death Cap, peels quite as easily as the mushroom. Colour also provides no safe guide: there are brightly coloured edible species like the Chanterelle and brightly coloured poisonous species like the Fly Agaric; there are cream-coloured or brown edible milkcaps and cream-coloured or brown poisonous milkcaps — both belonging to the genus *Lactarius* and not easily distinguished from each other by a novice. Neither do toothmarks and nibbled portions of a toadstool guarantee its safety, for the mice and rabbits that have eaten it have different digestive systems to humans and may be able to break down a toxin to safe derivatives. The only way to be sure is to have the toadstools identified by an expert: there are fungal forays in the autumn, run by natural history societies and led by experts, in many parts of Britain.

Rather different from the toadstools and more easily identifiable are the puffballs, the best of which is the gigantic (in fungal terms) fruiting body of *Lycoperdon giganteum*. It looks like an extra large football, up to 80 centimetres (30 inches) across, and grows in woodlands or pastures in the summer or autumn, during which time it produces over 10,000,000,000,000 spores. Presumably, since these Giant Puffballs are much less common than the smaller Common Puffball (*L. perlatum*, plate 7), very few of these spores survive to produce a mature fungus. The puffballs are not only used in cooking — they are eaten while still young and tender, when they have a cheesy consistency — but also in herbal medicine, when they are dried and used as a styptic to stop bleeding.

Truffles are perhaps the most famous of the fungal delicacies. They grow completely underground in deciduous woodland, where they form rounded, often warty, fruiting bodies in the late summer and autumn. For many centuries gourmets and scientists alike were baffled by these structures, not recognising their fungal nature and thinking them 'concretions of earth' instead. When 'ripe' the truffles emit a characteristic scent and can then be located by trained dogs or by pigs, which consider them as great a delicacy as their trainers. There are several species of truffles, not all edible, and most common in the warmer parts of Europe. The commercial truffle is the Black Truffle (*Tuber melanospermum*, fig. 10), which grows in France; only one edible species is at all common in Britain, the Summer Truffle (*T. aestivum*), which looks like a smaller version of the Black Truffle. It grows beneath beech trees, particularly on chalky soils, and is thought to have been introduced from France in the seventeenth century with imported trees.

The ecology of fungi

Woodlands provide one of the best places to see fungi, particularly in the autumn, when most of them produce fruiting bodies, toadstools, brackets, cup-shaped apothecia, stinkhorns, and so on. They are most common in the deeply shaded areas, typical of beech or coniferous woods for instance, where the soil is too acid and the light too dim to support the life of green plants, where leaves or needles cover the ground and provide a rich food supply for the mycelium of the fungus, and where the mycelial strands do not have to compete with the roots of such vigorous plants as Dog's Mercury or Enchanter's Nightshade. Coniferous woods are particularly rich in fungi, including species of *Boletus*, *Lactarius*, plate 15), *Clavaria*, *Russula* (plate 16), *Hydnum* and *Tricholoma*. Other woodland fungi, including species of *Amanita*, *Russula*, *Lactarius*, *Boletus* and *Peziza*, grow in mixed deciduous woods, often associated with certain species of deciduous trees. Beechwoods, for instance,

Fig. 10. *The most valuable of the truffles is the Black or Perigord Truffle (*Tuber melanosporum*), which is found in many parts of France. It grows beneath oak trees, where it can sometimes be detected by the swarms of truffle flies hovering above it, attracted by the scent. These flies lay their eggs in the truffles. It grows up to 8 cm in diameter, up to three times as large as the average English Summer Truffle (*T. aestivum*), which is otherwise similar in appearance.*

Plate 16 (left). *Russula atropurpurea has a dark purple cap, up to 10 cm across, and creamy white gills. It grows in summer and autumn, usually under oak or beech trees. Like many of the others, this Russula is edible but all are acrid when raw and have to be cooked before being eaten. The Russulas are common toadstools of woodland areas, readily distinguished from other genera by their white or cream-coloured rigid fragile gills and brittle flesh, which breaks easily into fragments.*

Plate 17 (right). *Toadstools of Mycena leucogala growing on pine needles; its deeply grooved caps are not more than 2 cm in diameter. This is one of many very small clump-forming Mycena species to be found in the leaf litter of deciduous woods, amongst pine needles or on decaying tree stumps in the autumn.*

Plate 18. *The Earth Fan (Thelephora terrestris) forms tough encrusting fan-shaped or rosette-shaped fruiting bodies in the litter of pine woods, on twigs and pieces of dead wood. It grows up to about 6 cm across, in late summer and autumn, especially in areas with light sandy soil.*

Plate 19 (right). *Stereum hirsutum causes a destructive white rot when it grows on posts or on stored hardwoods, but in the wild it grows on dead branches and stumps of deciduous trees like beech and oak. It is very common in deciduous woods and can be found throughout the year, especially in a wet season. It forms many overlapping, partly encrusting fruiting bodies, each up to 10 cm across, pale creamy yellow on the lower porous surface and concentrically zoned in tones of grey on the upper.*

Plate 20 (left). *The Yellow Antler Fungus (Calocera viscosa) is a common member of the Club fungi, basidiomycetes with club-like or coral-like fruiting bodies. Its fruiting bodies, 10 cm high, are tough but rather gelatinous and appear on the stumps of old conifers in autumn.*

Plate 21. *Clumps of golden-brown scaly toadstools of Gymnopilus junonius are a common sight in late summer and autumn in deciduous woodland, growing on tree stumps and at the bases of trees. The individual toadstools are quite large, their caps reaching up to 15 cm in diameter.*

are rich in fungi, with many species of *Cortinarius*, *Russula* and *Marasmius*, but oakwoods have far fewer, mostly *Russula* and *Boletus* species. Birch trees are often found with the brilliant red toadstools of *Amanita muscaria*, the Fly Agaric, or the Brown Roll-rims *(Paxillus involutus)* growing beneath them (plates 4 and 12).

The relationships between the trees and the fungi are complex and often species-dependent. In the simplest relationships the fungi live as saprophytes in the leaf litter, breaking down the nutrients in the leaves into simple chemicals which dissolve in the soil water to be absorbed by the roots of the trees. Some species are associated with particular fragments of leaf litter (plates 17 and 18): for instance, several species of *Mycena* grow amongst pine needles and on fallen twigs in coniferous woods; *Thelephora terrestris*, the Earth Fan, also grows on fallen pine twigs, while *Auriscalpium vulgare* forms small toadstools on decaying pine cones. Rotting stumps and logs provide habitats for many different kinds of fungi: encrusting forms like *Stereum hirsutum* (plate 19); cup fungi like the Scarlet Elf Cup *(Sarcoscypha coccinea)*; and club fungi like the Yellow Antler Fungus *(Calocera viscosa,* plate 20). Larger toadstools found on such rotting wood include *Gymnopilus junonius* (plate 21) and the striking yellow clumps of *Hypholoma fasciculare,* the Sulphur Tuft, illustrated on the front cover; brackets, such as *Coriolus versicolor* (plate 22) are also commonly found. Many basidiomycetes form mycorrhizal associations with the trees, that is to say the mycelia of the fungi become closely associated with the roots, penetrating into them, sheathing them in hyphae and apparently helping the roots to absorb water and nutrients, particularly phosphates. It is very probable that trees like pine and birch, which often grow in poor acid soils, are able to do so because, with their root mycorrhizas, they can absorb nutrients more effectively than they could with their roots alone. Often there are several species of mycorrhizal fungi associated with one species of tree; for instance growing with *Pinus sylvestris*, the Scots Pine, are *Amanita muscaria, Cortinarius mucosus, Lactarius deliciosus, Russula fragilis* and several species of *Boletus*. Conversely one fungus species may form mycorrhizal associations with several trees: *Amanita muscaria*, the Fly Agaric, for instance, grows with *Picea abies* (the Norwegian Spruce) and *Larix europaea* (the European Larch) as well as with the Scots Pine; however, it is mostly found growing with the Silver and Downy Birches *(Betula pendula* and *B. pubescens)*.

Mycorrhizal associations are probably far more common than is generally realised. It has long been known that orchids are dependent on mycorrhiza for proper root development, and that heathers cannot grow properly unless they form mycorrhizal associations, but it has now become evident that many other plants also grow more vigorously if associated with fungi, including cereals and many other crop plants. Unfortunately treatment with fungicides for disease-causing fungi such as grey moulds or Damping-off Disease also kills essential soil fungi and mycorrhiza, leaving plants less vigorous and their roots without protection. They are then more susceptible to insect attack, and to bacteria and other fungi, necessitating further spraying. In a very short while the whole balanced ecology of the soil can be upset.

Grassy pastures and woodland rides have fewer species of fungi than the woodland areas for a variety of reasons. There is more competition from green plants, such areas are more exposed to the wind and the humidity is lower — all factors which inhibit the growth of the fungi. However, the small, brightly coloured toadstools of *Hygrocybe* are often to be found, along with species of *Lepiota* (plate 14), *Tricholoma* (including the edible St George's Mushroom, *T. gambosum)* and *Agaricus* (including *A. campestris,* fig. 6, and *A. arvensis*. Many of the grassland toadstools grow in rings, like the Horse Mushroom *(A. arvensis)* or the Fairy Ring Toadstool *(Marasmius oreades,* fig. 11).

Cow pats and heaps of horse droppings, common sights in pastures and on bridle paths, provide small damp homes for a specialised assortment of fungi which help to decay the dung, releasing nutrients and fertilising the ground in the

Fig. 11. *Fairy Ring Toadstool (Marasmius oreades), showing fairy ring (below) and clump of toadstools (above). Individual toadstools are quite small, the caps reaching only 5 cm across. Despite their place in folklore, there is a natural explanation for the circles. The fungus starts off as a small patch of mycelium in the soil and gradually spreads outwards, releasing nutrients in the soil at its actively growing edge and creating a ring of lush grass in the process. Just behind this zone, the mycelium is at its densest and the grass is killed or stunted, creating a ring of bare ground. Further inside the ring the old mycelium dies and the grass becomes re-established. The ring may grow up to 60 cm a year and old rings may be up to 50 metres across.*

process. The fruiting bodies of these fungi appear in a very definite order, beginning with phycomycete moulds like the *Pilobolus* species (fig. 12). These moulds produce very beautiful translucent fruiting bodies tipped with black sporangia, which are ejected violently away from the stalk to land some distance away from the dung. If eaten by cows or horses cropping the grass, they pass through the intestines unharmed and are present in the dung when it is deposited. Most of the dung fungi have these characteristics in common, that the spores are flung or blown away from the dung to land on clean grass, where they are more likely to be eaten, and that they do not germinate until they have passed through the intestines of a grazing animal.

The *Pilobolus* fruiting bodies are followed by those of many small ascomycetes, including the cup-shaped apothecia of *Peziza* and the flask-shaped perithecia of *Sordaria*, and then by the toadstools of basidiomycetes, many of them members of the genera *Panaeolus* (plate 23) and *Coprinus*, the Ink Caps. The small delicate toadstools usually grow in clumps

Plate 22. *The leathery fruiting bodies of* Coriolus versicolor *(formerly* Trametes versicolor*) form beautiful overlapping clusters on fallen logs and trunks of deciduous trees throughout the year. This is probably the most common of the bracket fungi in Britain.*

Plate 23. *Toadstools of the little* Panaeolus semiovatus, *growing on horse dung. They may be found throughout most of the year on dung in the later stages of decay. Their caps remain bell-shaped even when the toadstools are fully mature, never opening up like those of many smaller species and only reaching 6 cm in diameter.*

Plate 24. *Toadstools of the Honey Fungus (*Armillaria mellea*). This grows in living trees and dead stumps, parasitically on the former and saprophytically after it has killed them. It is one of the fungi responsible for the strange luminescence which can transform decaying stumps at night. The toadstools have caps up to 15 cm across, pale yellow-brown to dark brown, sometimes with a greenish tinge.*

and have dark-coloured gills, black spores and conical caps.

Mycorrhizal and saprophytic fungi are not the only fungi present in the soil of woods and meadows. Less welcome are the parasites, the fungal species capable of causing infection, disease and death in their plant or animal hosts. The mycorrhiza seem to protect the forest trees against infection by the many potentially parasitic fungi present in the soil, such as the Honey Fungus *(Armillaria mellea,* plate 24) or the highly destructive Root Fomes *(Heterobasidion annosum,* formerly *Fomes annosus),* a bracket fungus which infects the roots of coniferous trees. The presence of the sheath of mycorrhizal hyphae around the roots appears to create a barrier that the parasitic hyphae cannot penetrate. However, there are other parasites against which mycorrhiza provide no protection, those whose spores are spread by wind or by other agencies like beetles or wasps. Dutch Elm Disease, which has devastated the elm populations of the United States, Europe and Britain, is caused by the ascomycete fungus *Ceratocystis ulmi* and is spread by beetles belonging to the genus *Scolytus.* It is introduced into healthy trees by young beetles feeding in the spring on new bark, spreads rapidly and soon kills the tree.

Mature female beetles lay their eggs in channels which they eat out of the dead bark, and the grubs feed on the bark, creating characteristic branched brood chambers in the process. Fungus spores in these brood chambers infest the young beetles as they emerge from their pupae in the spring and are carried with them when they leave.

Fungus diseases are numerous and varied, affecting many kinds of plants. They include the phycomycete *Phytophthora infestans,* the fungus which caused the potato famine in mid nineteenth-century Ireland by destroying the potato crop; the *Botrytus* (Fungi Imperfecti) species which cause the grey moulds on strawberries and other soft fruits; the rust and smut diseases of cereals (figs. 13 and 14); the ascomycete *Taphrina bullata,* the Pear Leaf Blister disease of pear trees; and the basidiomycete *Stereum purpureum,* the notifiable Silver-leaf disease of plums, both diseases capable of considerable damage in orchards. *Armillaria mellea,* the Honey Fungus (plate 24), is probably one of the most dreaded of the disease-causing fungi, destroying not only garden trees and shrubs, but also doing much damage in tree plantations. This fungus can be identified by its groups of honey-brown toadstools in the autumn and by its black bootlace-like

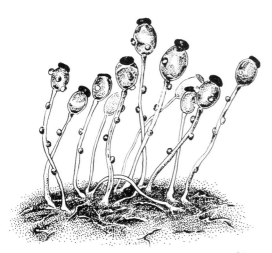

Fig. 12. *The sporangiophores (sporangium-bearing hyphae) of Pilobolus may be less than 0.5 cm high but have adapted to ensure their black sporangia are flung far outside the heap of dung on which they are growing. Each sporangiophore swells at the tip to form a water-filled bulb, so turgid that droplets of water are exuded on its surface. When the pressure becomes too great the bulb bursts just below the sporangium, which is then flung away as the water gushes out.*

Fig. 13 (left). *Ustilago avenae, Loose Smut on Oats*. One of the many fungi known as smuts (from the black colour of the spores), which attack cultivated cereals like oats, rye and wheat and their wild counterparts, infecting and destroying the flowers. Instead of producing grains, infected flowers produce large quantities of black spores which are blown by the wind into the soil, where they infect the young seedlings of the next crop.

Fig. 14 (right). *Puccinia graminis, Black Rust of Wheat*. This fungus disease can cause considerable damage to wheat crops. The name 'rust' comes from the rusty red colour of the spores. It forms two different kinds of spores on wheat leaves during the summer. The later ones are resistant spores which can survive the winter in the soil, germinating in the spring to produce basidiospores which infect the Barberry, *Berberis vulgaris* (shown behind), once widely planted for its fruit. This practice was discontinued when its part in the Black Rust life cycle was recognised. When the fungus is well established in the Barberry it produces spores which re-infect the wheat.

rhizomorphs (long strap-shaped structures consisting of many intertwining hyphae) spreading through the soil from an infected tree to others nearby. The fungus gains entry to its victim through the roots, spreads upwards into the trunk and eventually forms a mass of rhizomorphs encircling the wood to strangle the tree. It is more or less untreatable and affected trees are usually burned.

It is not only plants that are parasitised by fungi: a whole range of insects are hosts to the parasitic fungi of the genera *Cordyceps* and *Beauvaria*. Flies parasitised by *Entomophthora muscae* may be found staggering and buzzing on windowsills in autumn, dying within a few days to become mummified and surrounded by a circle of white spores, a danger to other flies investigating their dead companion. Fungi may also be predators, like those soil fungi that feed on eelworms by trapping them in special loop-like snares on their hyphae and then growing into the worms. One of the strangest associations between animals and fungi is that which exists between the Leaf-cutting or Parasol Ants of South America and certain species of fungi which the ants cultivate in their nests. The ants carry cut portions of leaves to their nests, chew them into pulp and incorporate them into underground gardens, where they are soon penetrated by the fungal hyphae already present. The fungi produce special spore-like structures on which the ants feed and pellets of fungus are carried by young queens on their nuptial flights, so that they can start fungus gardens of their own in their new nests. Ambrosia beetles cultivate similar fungus gardens in the chambers that they excavate in rotting wood, feeding on the spores produced by the fungi.

Economically important fungi

In many ways fungi play direct or indirect roles in our economy, directly as disease-causing parasites which threaten our crops, and indirectly in their roles as saprophytic organisms essential to the balance of nature and the annual cycles of growth and decay. But certain of them also play specialised roles in the food and drink industries and in medicine. Species of *Penicillium* and other moulds, for instance, have been used for hundreds of years in the making of blue cheeses. Blue-cheese making must have been rather a hit or miss operation when it originated in France. The cheeses were made from natural milk, full of contaminants, bacteria and fungi of all kinds, and left to mature in cool humid caves. The results must have been variable, depending on which bacteria or moulds proliferated in each cheese, but since the conditions in the cave did favour the growth of one particular mould, later named *Penicillium roquefortii*, the familiar Roquefort cheese was the most usual result. This mould is still used extensively today, in the making of blue cheeses all over the world, variations in texture and flavour depending on differences in salt content, production techniques and ripening conditions. Soft cheeses, like Brie and Camembert, are also produced with the aid of fungi, but in these cheeses the ripening and flavouring are induced by two separate fungi, *Penicillium camemberti* and *Oidium lactis*, which are introduced into the cheese at different stages of its production.

Yeasts of the species *Saccharomyces cerevisiae* are used extensively in the food and drink industries, in the making of bread and in the manufacture of wine and beer. The two processes utilise two different aspects of the yeasts' physiology. In breadmaking the yeast grows in the presence of oxygen and so respires aerobically, breaking down the sugar present into water and carbon dioxide gas. It is this gas which lightens the dough, creating pockets of gas in it and making it rise. In wine or beer making the yeast grows in a liquid (formed from grape juice or hops and barley respectively) which contains no oxygen, and so it respires anaerobically (without oxygen). This process is called fermentation and under these conditions the sugar in the liquid is broken down to alcohol and carbon dioxide gas, which is given off as bubbles.

Perhaps the most famous contribution that fungi have made to modern society is that of penicillin. This, still the most effective of the antibiotics, was discovered by Sir Alexander Fleming in 1928, when a culture of *Staphylococcus aureus*, a disease-causing bacterium on which he was working, was accidentally contaminated by the mould *Penicillium notatum*. Fleming noticed that there was a ring of dead bacteria around the mould. The significance of penicillin did not become fully apparent until the Second World War, when its effect on septic wounds and gas gangrene seemed little short of miraculous compared with the death and devastation caused by similar wounds in the First World War. Penicillin was also found to be effective against diphtheria, bacterial pneumonia, syphilis, gonorrhea and bacterial meningitis — all illnesses caused by gram-positive bacteria. During the war, large-scale research projects resulted in the production of mutant strains of *Penicillium chrysogenum*, strains which produced up to one thousand times as much penicillin as the original Fleming culture. At the same time equipment was designed so that full-scale commercial production could be implemented. Since then it has been discovered that there are several kinds of natural penicillin, each varying in its effect on the different illnesses. In addition several artificial penicillins are now manufactured from the natural substances, some like ampicillin active against gram-negative bacteria as well as against gram-positive ones, while others are active against bacteria which have become resistant to the natural penicillins.

Other antibiotics have been found since the discovery of penicillin, although most of them have come from the bacteria rather than from the fungi. However, streptomycin and other related antibiotics were isolated from the genus of ascomycete fungi *Streptomyces*, beginning with streptomycin itself from *S. griseus* in 1944, chloramphenicol from *S. venezualae* in 1947 and aureomycin from *S. aureofaciens* in 1948.

Streptomycin is different from many other antibiotics for, unlike the others, it is effective not only in human and veterinary medicine, but also in agriculture as a treatment for Downy Mildew disease of hops, a disease caused by the phycomycete fungus *Pseudoperonospora humuli*. Parasitic fungi like this Downy Mildew can cause millions of pounds worth of damage to many different crops each year, if not controlled. Fungicides are effective against many of them and new disease-resistant crop varieties are constantly being developed by plant breeders. The old mercury and sulphur based fungicides have now mostly been replaced by organic and systemic fungicides, which are generally more effective. Seeds are treated with these chemicals to prevent Damping-off Disease and trees and crops are sprayed to combat a variety of diseases like those described in the previous chapter.

Economically, it is not only the parasitic fungi which cause problems for the human population. In the wild the saprophytic fungi play an essential role in breaking down dead organic material, including wood, but in houses, huts and fence posts such fungi are very unwelcome. *Merulius lacrymans*, the Dry Rot Fungus, attacks untreated wood in all kinds of man-made structures, causing untold damage, especially in the damp cool climates of northern Europe and North America. Infection must begin in damp or wet wood, but the fungus can then spread into dry wood, producing enough water for its needs from the decay of the wood and even exuding drops of excess water. It spreads rapidly; its rhizomorphs grow through walls and between floors and its fruiting bodies produce millions of spores. Today structural timbers are usually treated against dry rot before being used in house building, as are external timbers, fence panels and sheds. Paint provides some protection for untreated wood if it is clean and dry before painting, but if damp or infected wood is painted, this can encourage decay by sealing the fungus in, together with the water which speeds its growth.

FURTHER READING

Hawksworth, D. L., Sutton, B. C., and Ainsworth, G. C. *Ainsworth and Bisby's Dictionary of Fungi.* Commonwealth Mycological Institute, 1983.

Ingold, C. T. *The Biology of the Fungi.* Hutchinson Educational Ltd, third edition 1973.

Kibby, G. *Mushrooms and Toadstools.* Oxford University Press, 1979.

Phillips, R. *Mushrooms and other Fungi of Great Britain and Europe.* Pan, 1981.

Ramsbottom, J. *Mushrooms and Toadstools.* Collins New Naturalist Series, 1953.

Webster, J. *Introduction to the Fungi.* Cambridge University Press, 1970.

ORGANISATIONS

British Mycological Society, Biodeterioration Centre, St Peter's College, University of Aston, Birmingham B8 3TE.

Commonwealth Mycological Institute, Ferry Lane, Kew, Richmond, Surrey.

The County Trusts for Nature Conservation often run fungal forays in the autumn. The addresses for the individual trusts can be obtained from the Royal Society for Nature Conservation, The Green, Nettleham, Lincoln LN2 2NR.

ACKNOWLEDGEMENTS

Photographs are acknowledged as follows: David Dimmock, plates 4, 12, 15, 17, 20, 22, 23; Keith Hayward, plate 13; Tony Mundell, plates 1, 2, 3, 5, 6, 7, 8, 9, 10, 11, 14, 16, 18, 19, 21, 24, fig. 11 and front cover. The line drawings are by P. Forey.